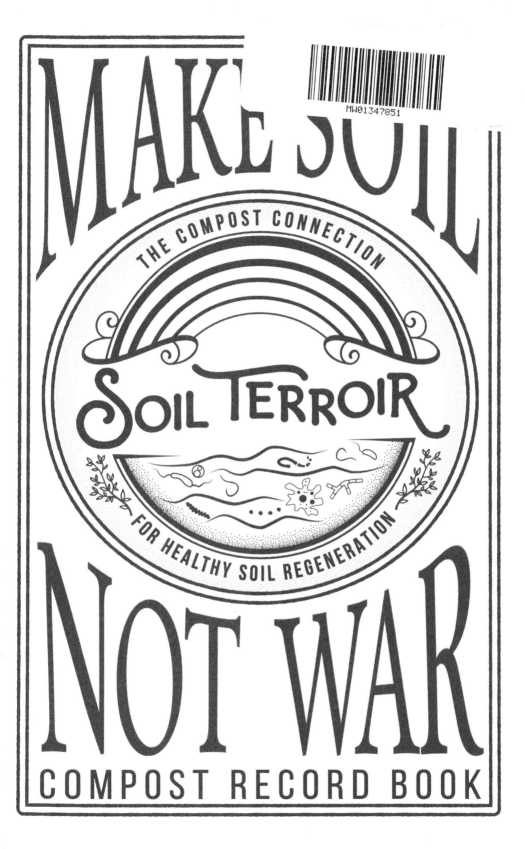

Copyright © 2022 Hilary Applegate, Soil Terroir, soilterroir.com
First Edition: March 2022

All rights reserved. No part of this publication may be reproduced, distributed, or transmitted in any form or by any means, including but not limited to photocopying, facsimiles or other electronic or mechanical methods, without the prior written permission from the designer, author or publisher, except where permitted by law.

Illustrations copyright © 2022 by Muñeca Osorio
Cover Art copyright © 2022 by Muñeca Osorio
Book Design by Muñeca Osorio, dangerouscat.com

ISBN-978-1-956590-03-6 (paperback)
Published by Moon Trine & Co. moontrine.com

Contents

Introduction 5
Basic Fundamentals... 6
Composting Basics 17
How To Use 30

The Compost Record Book - *Recording The Hot Compost Process*

    Pile #1 38-51
        Configure Recipe and Final Recipe
        Calendar View and Notes
        Daily Observation Log
        Compost Pile Summary Notes
        Draw Your Garden Map
        Lessons Learned and Where To Use Notes

    Pile #2 52
    Pile #3 66
    Pile #4 80
    Pile #5 94
    Pile #6 108

Appendix

    Static Pile Log 124
    Worm Bin Log 136
    Seasons Summary 140
    Q & A Notes 142
    Lists Of Educational References 144
    Educational Notes Workshops/Webinars 146
    Resources & Sourcing 150
    Additional Resources 154

Did you know
that no soil lacks
all the nutrients
a plant needs
to grow
and that a fully
healthy plant can be
completely resistant
to pests and diseases?

INTRODUCTION

This book will guide you to create and ensure an ecologically habitable environment for your plants, yourself, and others. You will be able to track your progress toward making a diverse, aerobic, microbe-rich compost. This is largely based on my training with the Soil Food Web School started by Dr. Elaine Ingham, a world renowned regenerative soil scientist. The school teaches how to make a pathogen-free, biologically complete compost and how to verify it with the microscope. They also teach how to make teas and extracts and how to apply them. This compost is organic matter that has been aerobically composted, and that meets the minimum biological requirements defined by scientific research and case studies. The studies demonstrate necessary plant-soil-human relationships (which many ancestors knew intuitively) that have yet to take a strong foothold in our fast-paced society.

We need people educating, practicing, and demanding these methods in order to push our institutions to adopt regenerative practices. This not only cleans up our environment, but it helps to grow strong people too. Dr. Ingham points to Dr. David Johnson's studies which show how it isn't N-P-K that correlates to the most growth in a plant. It is actually when the fungal to bacteria ratio (F:B) is highest that plants show the most growth and in turn exhibit their best health, yielding more productivity and nutrition. In this book we have an organized way to track making your own compost that targets optimum F:B ratios for the corresponding successional stage, or habitat, of your plants. With the Soil Food Web present, ecological functioning is restored for a myriad of environmental processes. Without it our soil literally crumbles and we fall prey to scarcity, rather than utilizing improved methods that can bring abundance.

In the past, people have been able to migrate to more fertile ground after desertifying it. In some cases, governments/corporations have taken or occupied lands where locals cannot sustain their community because of the extraction-based system. Increasingly, there are climate refugees who are forced to move because their environment is uninhabitable. This could be considered grounds for war. Blame gets placed without solving the *root* problem when we don't have a system based on interconnection which follows the regenerative laws of nature. Our centralized society has waged a war on nature using leftover chemical warfare to invent a whole industry to kill plants, the soil food web, and the pollinators, in the name of protecting it. Many know that this is treating symptoms and not the cause, so we need to make better solutions reasonable in these pivotal times. Composting is not a new concept. Thankfully, as environmentally conscious communities grow we can integrate enhanced compost techniques, which will hydrate the landscape and increase the nutrient density of the foods we eat. In light of more fertile ground I say **'Make Soil, Not War'**.

Hilary Applegate
Soil Terroir Owner & Soil Analyst

BASIC FUNDAMENTALS

Terroir [ter-wahr; French ter-war]

Terroir describes how the land from which plants are grown, together with growers' methods and practices, impart a unique quality that is specific to a habitat. Soil microbes are foundational to a habitat by allowing plants to thrive in their elements and by regulating our climate for clean air, water, and soil. Ensuring habitable soil is essential to all growing needs, therefore Soil Terroir.

Examples of a soil's terroir in creating valuable and unique qualities include:

Landscape
Nourish microbes locally to support the natural ecosystems of your area.

Food
Take care of your gut microbiome by supporting the soil microbiome.

Herbs
Provide nutrient dense oils for your health.

Wine
Grow to taste a wine's appellation.

As you develop your composting practice, you will learn ways to enhance microbe populations and diversity according to your plants' needs. This is accomplished by compost making techniques, and the number and types (compost, teas, and extracts) of applications to an area over time.

A Functioning Compost: What Is In It & Why?

Remember learning about photosynthesis? Decomposition with soil microbes is part 2 of that age old lesson.

Plants use solar energy to pull apart carbon from CO_2 molecules and with water make specific sugars (exudates) as energy to store down into their roots for connection, growth, and health. There is a reason plants expend so much of their energy to carry out this process. It is to attract and feed microbes that maintain the root zones, or rhizospheres, ensuring their roots are fed a proper balance of soluble nutrients.

Plants respond to their environment with functioning biogeochemical (life, mineral, water) cycles. Organic matter, water, air, and the composition of the life in the soil unlock and make available every nutrient that they need, causing decomposition and thereby making more soil and more plants.

Composting is a way we have used biomimicry to recycle organic material and create soil for our livelihood. This results in an accelerated rate of decomposition compared to the geological time-scale of nature. Most plants we depend on for sustenance need aerobic (with air) conditions for their roots to live in the soil. Applying aerobic microbe supported compost mimics the natural decomposition process for upland plants wherein they have a constant supply of nutrients precisely when they need it. This way other inputs aren't needed because the soil (containing microbes, minerals, organic matter, water, and air) is held in the landscape.

This compost recipe contains diverse, beneficial, aerobic organisms (6 parts per million of oxygen requirement) that will live and work in the rhizosphere, as well as cover all surfaces of the plant, once applied. Below 6ppm of O_2 supports a different set of organisms that will begin to decompose aerobic plants before their time. The soil becomes anaerobic and loses structure, which eventually transforms the landscape.

The goal in this compost record book is to make dark chocolate color compost piles containing diverse, beneficial, aerobic organisms that have at least:

- **135 µg/g** of Bacterial Biomass and **135 µg/g** of Fungal Biomass
- Fungal to Bacteria Ratio (F:B) of **0.3:1**
- **10,000/g** of Protozoa
- **100/g** of Beneficial Nematodes

(Source: Lecture Course Material, Soil Food Web School LLC, 2020.)

This is the minimum for a soil building, truly healthy compost and varies according to the successional stage of the plants you want to grow.

BASIC FUNDAMENTALS

Soil Food Web Trophic Levels

LEVEL 1
Plants

Organic Matter

LEVEL 2
Bacteria
Fungi
Root Feeding Nematodes

LEVEL 3
Arthropods (Shredders)
Protozoa (Amoeba, Flagellates, Ciliates)
Fungi & Bacterial Feeding Nematodes

LEVEL 4
Arthropods (Predators)
Nematodes (Predators)

LEVEL 5+
Birds (Predators)
Mammals (Predators)

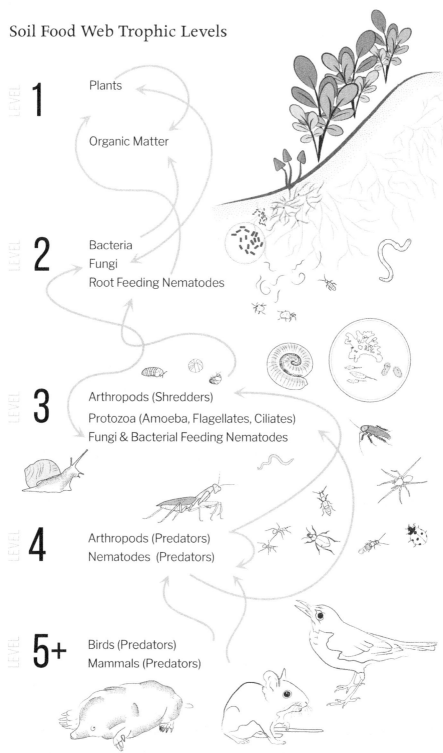

BASIC FUNDAMENTALS

Trophic: Relating to Feeding & Nutrition

By giving the right environment for the organisms to thrive, chemical processes occur that build well structured and highly fertile soil. When the quantities of different aerobic organisms are high enough, they can do their work to compete with, inhibit and consume the anaerobic organisms and maintain the aerobic environment in the compost and in the soil.

Plants express deep roots and tight nodes for higher yields and higher nutrient density. Again, plants expend a lot of energy to give off a diverse and specific set of exudates consisting of sugars, proteins and carbohydrates that attract certain species of bacteria and fungi. The bacteria and fungi feed off those exudates and also begin mining the parent soil for all the nutrients that plants need. Soil structure improves with the bacteria/fungi mining operation because it allows further natural chemical processes to form beneficial mineral rich aggregates. Bacteria mine the parent material, and organic material, with their alkaline enzymes. Fungi mine with their acidic enzymes. Therefore we make compost with food for each of them to break down so they will reproduce.

Then come the protozoa and the nematodes. The protozoa (the flagellates and amoebae) eat the bacteria. Different kinds of nematodes eat bacteria, fungi, or other smaller nematodes. Some nematodes eat roots. Those are the kind that show up in a not-so-aerobic environment, but if your numbers are high enough in the beneficial soil food web then you are covered. FUN FACT: There are nematode trapping fungi that form loops to trap nematodes and consume them!

The four major microbial soil food web groups to measure soil-plant health are bacteria, fungi, protozoa, and nematodes. They are directly within the rhizosphere and exhibit classic predator-prey cycles that maintain an ecosystem. For example, when bacteria are high, protozoa start increasing because there is ample food. Then bacteria numbers drop and then protozoa numbers drop, where again bacteria begin to increase and so on. The populations fluctuate up and down with different types of bacteria and protozoa depending on anaerobic or aerobic conditions. In tilled land there are simply no protozoa or fungi so we lose soil web function. This forces us to use greater quantities of fertilizers, herbicides and pesticides which cause desertification. It also forces us to use more land that could be wildlife habitat, whereby restoring our ecosystem.

Different species exist within the same and different kinds of habitats and regions, and between aerobic or anaerobic conditions. They cycle through the systems of water, air, soil, plants, and animals. The ones on our upland plants should be covered with aerobic types of organisms above and below ground so that nothing else can get to the plant and attack it. Supporting the populations of organisms at trophic levels 1, 2, and 3 means supporting the web of interactions among all the trophic levels. With this law of nature in mind, we can make decisions accordingly.

BASIC FUNDAMENTALS

Four Groups Of Micro-organisms

What makes it a 'micro' kind of organism? A micrometer(μm) is 1000 x's smaller than a millimeter, which is 1000 x's smaller than a meter. So organisms that are micrometers in size, need a microscope to see and/or measure them!

Meet the base level microscopic life residing in the soil:

Bacteria

Fungi

Protozoa

Nematodes

These are four health indicator groups that live in the soil biome.

The plants feed them and they feed the plants,
according to who's keeping who in check and in what conditions.

Beneficial Biological Life

With these four major groups all 5 kingdoms* are accounted for within the root and shoot systems of plants. The Plants being the conduit for connecting everything above and below ground. They are the lungs of the earth. The soil biome is like our gut biome, and the water is like our blood.

We look at these groups in the soil because they are key indicators for soil respiration, or health. We know that in each of these groups morphologies exist in anaerobic conditions and in aerobic conditions. We also know that the aerobes serve the functions for nutrient cycling at the root zone. We don't have to speciate each one for our purposes. We want diversity in the right conditions, so we are identifying the indicator groups for anaerobic vs. aerobic conditions.

We can see differences between the types of groups by:

Bacteria — their size, shape, form and movement

Fungi — their size, shape, form, and color

Protozoa — their size, shape, form, and motility

Nematodes — their digestive system, mouth parts, and size

As with all of them, the last way we identify their functions is what they eat and who eats them along the successional stages of plant groups.

FUN FACT

The 4 groups plus the Plants represent all 5 kingdoms on the planet, and they are all in the rhizosphere!

Together, our kingdoms: Fungi, Monera (the umbrella for bacteria), Protists (where protozoa are 1 of 5 major protists groups), Animals (which include nematodes) and our beloved Plants give us full ecosystem functionality.

---

*Kingdom (biology), Wikipedia, 2022

Key Indicators of *Anaerobic* Conditions:

Bacteria

Actinobacteria
Clear, filamentous strand at 1μm wide, flourishes in reduced O2 conditions and is beneficial for cruciferous plants, however prevents mycorrhizal colonization

Spirilla, Spirochetes, Vibrios and more
Bacteria in a chain or ones that are not rod or circular shape can be pathogenic

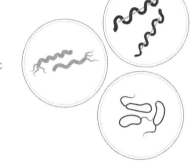

Protozoa

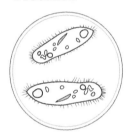

Ciliates
Very fast, usually larger than amoebae, have cilia around the perimeter of their bodies

Fungi

Oomycetes
Fuzzy looking with the naked eye, non-uniform, uneven septate, colorless, 1.5-2.5 μm wide

Nematodes

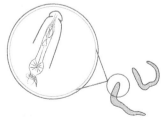

Root Feeding Nematodes
Distinguished by a knob near the mouth area

BASIC FUNDAMENTALS

Key Indicators of *Aerobic* Conditions:

Bacteria
Counted, then converted as a mass, biomass as μg/g

Coccus bacteria
Smallest are 1μm in diameter

Rod-shaped bacillus

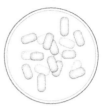

Protozoa

Flagellates
3-10 μm, round banana or pear shape, smaller than amoebae, have one or more flagella, rolling, bumbling motion

Amoebae (plural) / Amoeba (singular)
Slow, oozing movement, usually larger than flagellates but smaller than ciliates, with testates or without

Some are completely round with or without testates; Arcella are large donut-shaped testate amoebae.

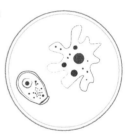

Fungi

Deuteromycetes, Ascomycetes, Basidiomycetes
Uniform in diameter, even spaced septate, 2-5+ μm wide, usually have color

Beneficial Nematodes

Predatory
Larger body with big tooth

Bacterial Feeder
Cylinder chamber behind fancy lips

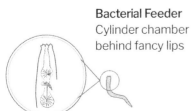

Fungal Feeder
Spear to puncture cell wall of fungi

BASIC FUNDAMENTALS

Living microbes are eating, re-producing, excreting *and moving* in response to their environment. Nature's operating system contains a biotic pump where plants respire micro-organisms to form nuclei for rain clouds.

People can facilitate this by creating small-scale earthworks, planting plants, and applying compost that will help water to recycle through and around the landscape. This restores local water cycling, known as the small-water cycle, accounting for most of a region's rain. *(WaterStories.com, Desert or Rainforest with Walter Jenhe, 2022)*

A plant's successional stage shows the biological composition they require. Applying a compost targeting the plants you desire can also serve to detoxify our water, soil, and air by restoring elemental processes such as the nitrogen, carbon, phosphorus, and sulfur cycles.

What was the historical use of your growing space? What is the natural environment? Which successional stage is your growing space in currently? Where does it need to be to prevent unwanted plants?

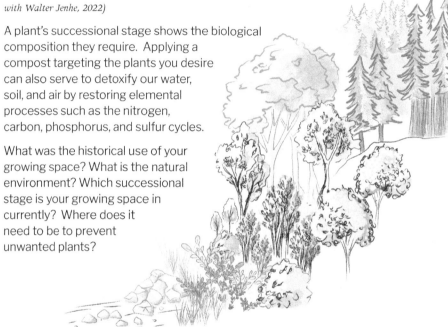

Bare, Cyanobacteria, Exposed Rocks	Lichen & Mosses	Annuals	Mid Stage Grasses / Vegetables / Row Crops	Perennials Herbaceous Plants / Shrubs	Vines/Bushes	Deciduous	Conifer
F:B 0.0	.01	.1	.75 / 1:1	2:1	5:1	100:1	1000:1
100% Bacterial		Protozoa	Bacterial Feeding Nematodes	Fungal Feeding Nematodes	Predatory Nematodes		Micro-Arthropods
NO3 (Nitrate)			Balanced NO3 and NH4				NH4 (Ammonium)

DEFINITIONS

SUCCESSION:
Process describing how the structure of a biological community changes over time.

BIOMASS:
Total mass of organisms in a given area or volume.

BASIC FUNDAMENTALS

Stages of Succession
RATIOS OF ORGANISM BIOMASS THROUGHOUT SOIL-PLANT SUCCESSION

SOIL-PLANT SUCCESSION STAGE	FUNGI:BACTERIA RATIO (F:B)	PREDATORS
Dirt No vegetative cover, lack of structure, easily compacts	F: <5 µg/g B: <45 µg/g F:B <0.01	None No nutrient cycling
Bare Soil No veg. cover, poor structure, easily compacts, nitrate cycling only	F: <23 µg/g (disease type likely) B: >450 µg/g F:B ~ 0.05	**Protozoa:** Ciliates **Nematodes:** Rare Anaerobic conditions causing loss of nutrients
Weeds Nitrate only, shallow depth with compaction except tap root type later succession, little soil building	F: <45 µg/g (disease type likely) B: >450 µg/g F:B ~ 0.1	**Protozoa:** Ciliates **Nematodes:** Bacterial-feeders Nutrients pulse low and high
Early Successional Wetlands, Brassicas. Shallow root depth except for tap roots providing food and breaking compaction	F: 32-90 µg/g (mostly disease type, hardly if any mycorrhizal) B: >135-270 µg/g Actinobacteria protection from mycorrhizal colonization F:B ~ 0.3	**Protozoa:** Flagellates and amoebae ~10,000/g soil; normal nutrient cycling **Nematodes:** Bacterial feeders present nitrate high, ammonium low
Vegetables & Early Successional Grasses Root-crops, lettuce, greens, Bromus, Bermuda etc.; nitrate and ammonium needed	F: 68-225 µg/g, most species require mycorrhizal colonization (VAM) B: 135-450 µg/g F:B ~0.5	**Protozoa:** Flagellates and amoebae 10,000-50,000/g soil during growing season **Nematodes:** Bacterial feeders, fungal feeders, predatory
Mid-successional Turf (I.e. Ryegrass), Vegetables, Annual Crops And Flowers Still predominately nitrate, needs ammonium	F: 101-1012 µg/g B: 135-1350 VAM required F:B ~0.75	**Protozoa:** Flagellates and amoebae > 50,000µg/g soil in growing season **Nematodes:** Bacterial feeders, fungal feeders, predatory
Productive Pastures, Row Crops Lawns w/o Weeds, Corn, Wheat, Barley, etc. Equal nitrate and ammonium	F: 135-1350 µg/g B: 135-1350 µg/g VAM required F:B~1.0	**Protozoa:** Flagellates and amoebae > 50,000/g during growing season Nematodes: Bacterial feeders, fungal feeders, predatory necessary
Shrubs, Bushes, Vines Require more ammonium than nitrate; fungal greater than bacterial	F: 270-6750 µg/g VAM / Ecto / Ericoid required B: 135-1350 µg/g F:B ~ 2.0-5.0	**Protozoa:** Flagellates and amoebae > 50,000/g in growing season **Nematodes:** Fungal and predatory, microarthropods rivaling bacterial feeders
Deciduous Trees Require mostly ammonium; nitrate encourages disease fungi	F: 675-9,000 µg/g; VAM required B: 135-900 µg/g F:B ~ 5.0-10.0	**Protozoa:** Flagellates and amoebae > 10,000 in growing season **Nematodes:** Fungal and predatory should equal bacterial, unless function is replaced by micro
Conifer/Evergreen Trees Just before the growing season begins, most of the weight of a gram of soil will be fungal; Require strictly ammonium; nitrate will harm trees	F: 1,350-45,000+ µg/g; Ecto required B: 135-450 µg/g F:B >10.0	**Protozoa:** Not as important in a fungal dominated system; > 10,000/g soil in growing season. **Nematodes:** Fungal and predatory nematodes should exceed bacterial numbers unless their function is replaced by microarthropods

Source: Lecture Course Material, Soil Food Web School LLC, 2020.

BASIC FUNDAMENTALS

Feeding Microbes

All living organisms have a set carbon to nitrogen ratio (C:N) and when they consume another being that has a different C:N ratio, they excrete the excess to maintain their C:N ratio. That excess is one of the ways that plants get soluble nutrients in the rhizosphere. Different types of beneficial fungi provide nutrients *and water* to plants in various ways throughout their life cycles. Composting microbes need carbon for energy and nitrogen for protein production.

Ingredient proportions for the Hot Compost Recipe:

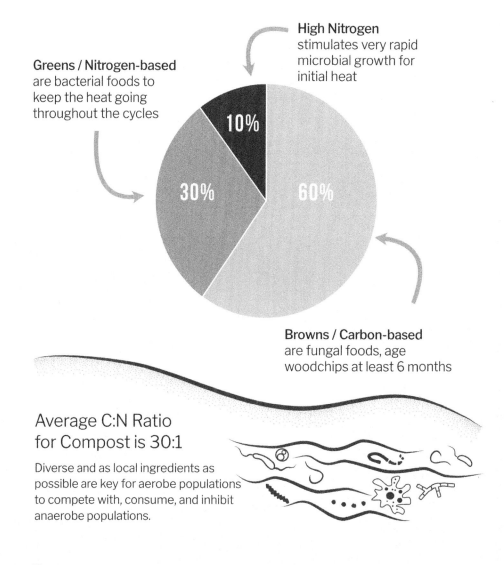

High Nitrogen stimulates very rapid microbial growth for initial heat

Greens / Nitrogen-based are bacterial foods to keep the heat going throughout the cycles

Browns / Carbon-based are fungal foods, age woodchips at least 6 months

Average C:N Ratio for Compost is 30:1

Diverse and as local ingredients as possible are key for aerobe populations to compete with, consume, and inhibit anaerobe populations.

COMPOSTING BASICS

Different Methods Of Composting
Hot Pile vs. Static Pile vs. Worm Bag

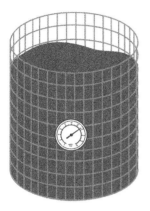

With a **Hot Pile**, the thermophilic process takes around 25 days, from start to cool down. All ingredients are mixed together at the start. It has to reach a certain temperature for a certain duration in order to kill pathogens, unwanted plants and their seeds. Because this heat only gets up to temperature in the middle, the pile must be turned a certain way so that each part goes through the thermophilic process. Air is also let in by turning so this keeps it an aerobic process.

A **Static Pile** on the other hand, doesn't account for temperature durations, and does not get turned or opened up until at least 2 weeks after you stop adding to it. At this point you could leave it and let nature keep breaking it down or thermophilically compost it or add to a worm bag.

Vermicomposting, the worm bag method shown here, is a cool temperature composting process. Worms do an immaculate job of killing pathogens but do not get rid of unwanted seeds. It is a superior way to process food waste. However, it takes time and the right environment to increase the amount of food waste.

COMPOSTING BASICS

Useful Tools For Hot Composting

Rubber Gloves
Up to elbow length to check moisture down into pile and eyeing droplets on your glove as a way to measure moisture

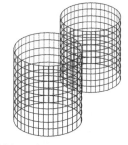

3' Soil Thermometer
With temperature range 0° to 200° F

2 Compost Vessels
Hardware cloth fastened together with string into a cylinder shape

2 Pallets and Screen Material
Find free pallets from a garden/hardware or feed store, lay screen material over each

Wheelbarrow
For hauling and/or measuring ingredients (count how many 5-gal buckets it will take)

5-gallon Buckets
40 buckets for making up to a 200 gallon pile. Tractor Supply Co. has a good deal, or find free at grocery stores/restaurants but will take time to collect all you need.

Old Sheet
Get from the thrift store if you don't have one. This is to cover piles after they are done.

Tarp
Use for covering piles from rain. Pitch it like a tent over the compost (stick stuck into it and drape tarp over). If windy, secure it down. Also used for turning piles.

PitchFork
For mixing ingredients and turning the pile

Clean Filtered Water
If you don't have rain catchment water to use, get an inline hose filter that filters out Chloramine and Chlorine.

Spray Nozzle Attachment
Get a 'Fog-it' attachment for a very fine mist and/or a wand that has different settings.

Hot Pile, aka Thermophilic, aka 'Compost Cake'

Thermophilic means organisms living at a high temperature. In a thermophilic compost pile aerobic organisms respire and populate very fast. Being fed the right ingredients with the right moisture and oxygen causes big jumps in heat. This happens in the center of the pile and we turn it so that each part (top, center, and bottom) has a chance to go through the heat process. This mandates 3 heat cycles because it has to be turned after the first and second heat cycles to make sure all parts of the pile get treated in the hot middle. (See pg. 21 for how to turn.) This will kill pathogens, unwanted plants, and seeds, and produce good populations of diverse, aerobic, and beneficial organisms.

One heat cycle consists of any one of three possible temperature/time thresholds and corresponding time durations. These thresholds are: 131° for 72 hours, 150° for 48 hours, 160° for 24 hours. If your pile reaches 170° turn immediately to lower the temperature and avoid over cooking your pile (or having it spontaneously combust). Your thresholds can change depending on meeting those temperatures at any point in time during each cycle.

For example, if the temperature meets 131° then you start your thermophilic process clock, and if it stays under 150° then it will take a full 72 hours until the heat cycle is done. But if it reaches 150° then it takes a shorter amount of time; instead of taking 72 hours total it will take 48 hours total, the same for if it reaches 160°, then it only takes 24 hours.

Once the time is up for that temperature/time threshold, the pile must be turned as soon as possible to avoid anaerobic conditions building up. As heat increases the aerobic organisms could use up too much oxygen too fast and die if it isn't turned and aerated. This is a good time to add water as the pile needs to be maintained at 50% moisture throughout the hot process. The third heat period needs to finish to the end but doesn't have to be turned in the proper way because each part of the pile has already had a chance to be treated in the middle. It can be moved to a big Smart Pot, or on the ground and covered with an old sheet for storing.

Depending on the season, the pile will hold more moisture in cold weather and dry out more quickly in hot weather. Keep this in mind as it needs to be monitored for its moisture content to remain at 35%. (See pg. 33 for how to check moisture.)

Ingredients For Making 'Compost Cake'

Using at least 120 gallons of material, it is best to have diverse ingredients with 3 kinds or more of each of these categories. Obtain wood chips, manure, weeds, grass clippings and any material you can from your local area. The microbes within these materials have evolved in the local climate and soil biome for millennia so will be better adapted to perform their work to compete with, consume, and inhibit the anaerobic populations. Let woodchips age for at least 6 months so their microbe inhibiting oils dissipate.

10% HIGH NITROGEN	C:N	30% GREEN MATERIALS	C:N	60% BROWN MATERIALS	C:N
Average Range	5-15:1	**Average Range**	20-30:1	**Average Range**	
Spent Brewer's Yeast		Finished Compost		Wood Chips	500:1
Cracked Corn		Food Scraps		Cardboard	400:1
Bird Seed Mix		Garden Plants		Brown Leaves/Needles	150:1
Meat, Fat, Blood Meal		Used Coffee Grounds		Paper Products	120:1
Chicken, Pig Manure		Grass Clippings		Straw	75:1
Horse, Cow Manure		Alfalfa - Green			

Nitrogen & Carbon

Have at least 2% of your high nitrogen be a consistent material, such as bird seed, cracked corn, and/or spent brewer's yeast. Most manures, unless the animals are fed with optimum nutrients and the manure is uberly fresh, do not have enough nitrogen to raise the initial heat needed. Depending on the time of year and the size of the pile you can play with different ingredients for carbon to nitrogen ratios, but keep the 10, 30, 60 rule. If the pile fails (does not reach temperature/time threshold due to insufficient nitrogen), you can use the remains as woody ingredients for your next pile because it's still good, diverse material. Carbon/fungal foods take longer to break down so it takes fungi longer to reproduce. Prepare them wisely to get a higher F:B ratio.

Moisture & Preservation

All materials need to start at 50% moisture and/or once mixed it needs to be about 50%. Kitchen waste has about 100% moisture and green garden trimmings have about 80% moisture content. Any dry materials, like woodchips, bird seed or dried grasses, need to be soaked for at least 24 hours. Soaking woody ingredients for 2 days or longer in water or in extract is beneficial to grow more fungi in your pile.

There is also a way to preserve your fresh ingredients to save for later which requires drying, soaking, and draining. *If you aren't going to use your fresh manure, grass trimmings, garden weeds etc. within 24 hours then leave them out to dry for about 24 hours. Then collect into buckets with a lid. When ready to use, soak for 24 hours and drain any excess water (to avoid making the pile too wet) before adding to the pile.*

COMPOSTING BASICS

Initial Set-up Of The 'Compost Cake'

Have your space ready with two pallets, each covered with screen material, and two compost vessels plus an area to put ingredients on a sheet.

Use the Configuring Scratch Recipe page in the record book section to begin computing your materials. Record how you measure your ingredients, a suggestion is to measure by the gallon and use 5-gallon buckets. Eyeball ingredients in the bucket for measuring ingredients that are less than 5-gallons.

Thoroughly mix all ingredients on an old sheet, or tarp, and then put these ingredients in your compost vessel. After each turn you have up to 24 hours for your pile to get back up to 131°. If it doesn't, it allows pathogens to recover. After you turn a second time and complete the 3rd heat cycle, you can either leave the material in the vessel or transfer out onto the ground or into smart pots, as long as temperatures aren't too high and/or rising.

How To Turn Your Pile

Imagine your compost pile in 3 equal parts consisting of a top and bottom that encloses a big ball in the middle. Observe where on your compost vessel those sections are in relation to the pile inside:

Step 1 Using your pitchfork, take the top and put it on a tarp/sheet, set aside until step 3.

Step 2 Carving out the middle ball, place it into the bottom of the second empty compost vessel that sits on another screen-covered pallet. Form a little depression in the center for the new middle ball shape you'll add in the next step.

Step 3 Then using the compost that was set aside on a tarp/sheet, fill the middle part of the second compost vessel.

Step 4 Take the bottom from the first vessel and place it on top in the second vessel.

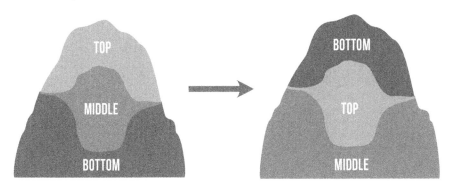

Temperature Thresholds & Time Durations

Temperatures are in Fahrenheit degrees.
Compost is ready after a minimum of 3 heat cycles.

1 heat cycle consists of a temp/time threshold of either:

- 131° to 149° maintained for 72 hours then turn
- 150° to 159° maintained for 48 hours then turn
- 160° to 169° maintained for 24 hours then turn

If the temperature increases at any point in time to the next threshold, the heat cycle changes to that threshold's time duration as well.

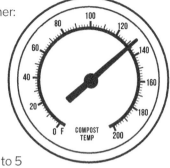

130° This period could take a few hours or up to 5 days in order to reach 130°. If it rises very quickly within a few hours but is not yet up to 131°, you can anticipate that your pile will be rising too fast overall and you'll be forced to turn it often. You can slow the temperature rising by adding cardboard and it will increase fungi growth. Conversely, if not continually rising, it may help to add ½ cup or more of blood meal into the center of the pile. If your pile doesn't heat up initially to start the first cycle or if it doesn't rise up to 131° within 24 hours after turning in any of the following heat cycles, the pile has failed. You can still use all that material as woody ingredients for your next hot pile.

131° Log the time you observe 131° for the 72 hour thermophilic process. This is the point where weeds, weed seeds and pathogen-causing organisms begin to die. If the pile maintains below 150°, then one heat cycle will complete after 72 hours.

140° Typically at 140°, normal anaerobic conditions begin to develop because oxygen is being used up by microbes and causes Actinobacteria (looks white/ashy) to proliferate. Monitor moisture levels, aim for 50%.

150° If the pile heats up to 150°, only 48 hours will be needed to complete one hot cycle instead of the 72 hours. Whichever time duration you are in, when you turn your pile it drops in temperature 10-15 degrees so aiming for 150° before turning is ideal. However, don't wait too long if it has already been over 131° for 72 hours, as it may get too anaerobic.

160° If the pile heats up to 160°, only 24 hours are needed to complete one hot cycle. After 24 hours, turn the compost pile.

COMPOSTING BASICS

When to Turn and/or Push Holes

170° If temperatures rise to 170° before the end of any heat cycle threshold, turn immediately and don't count it as a heat cycle. Once you've turned the pile, wait at least one hour before measuring temperature to determine the heat cycle threshold the pile is in and restart the cycle.

150°- 169° If temperatures are rising quickly you may poke holes all the way through the pile so you can maintain the temperature for the duration of the heat cycle you are working with. Consider the threshold cycles and time of day; anticipate having to turn immediately and/or prevent having to turn in the dark according to the heat cycle. (For example, if you take a reading at night and see the pile is increasing to 160°, punch holes to prevent turning in the dark the next night and to prevent temperatures from increasing to 170°+ overnight.) Consider a slim to wide punch tool (thermometer or broomstick) vs. amount of holes vs. temperature and heat cycle duration; can add holes gradually and monitor closely.

131°+ If temperatures are still over 131° + after several hours of completing the third heat cycle, aerate it by continuing to turn and/or punching holes in it. Monitor until the temperature stays below 131°. This means you have added too many greens and/or high nitrogen material and have lost a lot of nutrients in the gas blow-off. You will need to reconfigure your recipe for next time.

After All Heat Cycles Are Complete

At ambient temperature, the pile has stabilized and is ready to amend your soil. You could also order a compost test to examine the soil food web and determine your next steps. For example, performing a nematode extraction or a protozoan infusion can aid you in getting your pile more of what it needs. As well as, adding sterile fungal foods, like oats, after the pile is done increases beneficial fungi over time.

Ideally, this compost should make an average F:B ratio of 1:1. Time alone also increases fungal biomass. Although, do not wait too long to use your compost. After six months the pile will continue to decrease in size considerably and you miss out on applying it to your farm/garden.

COMPOSTING BASICS

Static Pile Compost

There is no hard science with a static pile, as there are many permutations to it, so just settling in to understanding how decomposition works and relating that to what you want to do in your garden can help you decide how, where and with what material you make your static compost pile. Like taking care of a pet, your pile needs food, shelter, and water.

One idea of a static pile is to utilize your kitchen scraps as pile deposits to reduce methane in the landfill. To prevent nutrient loss in your kitchen scraps, they should only be about 1 or 2 days old or frozen or dried out and stored before it goes out to the pile. If possible, it's best to have a container on your counter that you add to a bucket in the freezer until you are ready to deposit into your static compost pile. Another idea is adding a bit of woody material to your kitchen pail every time you add scraps until you take it out to the pile again.

Depending on the amount of kitchen scraps vs. how often you make deposits vs. how many deposits, the pile could be 3 stories high if you want! In that case you would need a front-end loader and a ladder for layering on the brown/green mix. You could dig into the ground and start your compost base in a hole so it's not so high. Consider earthworks (like a hügelkultur bed on contour), water flow, and what plants might get fed around the pile when choosing its location.

Each deposit will go anaerobic so there needs to be aerobic layers to absorb the gasses from the alcohols they produce. Decomposition by aerobic organisms will make compost in these layers so moisture needs to be maintained there. Moisture isn't checked the same way as the hot pile since the static pile doesn't get opened up, but watering as you go with the idea of not being too wet and not too dry should suffice. Covering with an old sheet for rainy days prevents compaction leading to anaerobic conditions, the sheet will also keep moisture in on sunny days.

In using the Static Pile Log (pages 122-133), you are keeping track of your pile with the idea that you will know basically what you did through the process so you can see what works and what doesn't. In the Static Pile Log you can keep track of when you started and stopped, what you added and when, and how often it receives water. You can then plan when you will thermophilically process it, vermicompost it or just let nature take care of it.

The static pile base layer deposit placements are shown here and are described in detail on the next page.

Example of a 14-week Static Compost Pile

Tools and Materials: Kitchen compost container, container for freezer (or dehydrate, take out more often and/or add browns to your kitchen scraps on counter), something for marking where the latest deposit is, and either a watering can, hose and a water filter, rainwater catchment, or freshwater source.

Ingredients: Stick-like woody material, wood chips mixed with grasses/weeds, kitchen scraps added over time, and a high nitrogen ingredient possibly later. Grasses and weeds can be dried for a day and stored, or used fresh. Make sure to use aged woody material that you have soaked for at least 24 hours.

First, build up a stick-like woody layer about 5' x 4' x 2' for the base. This can be smaller or larger but if you are going to hot compost it later you will need mass to keep the heat throughout all heat cycles. Next, top it with about a half foot layer over that of mixed smaller woody and green type ingredients such as weeds or grass trimmings or dried greens from a feed store.

If you do not have green material just use woody material, but be sure to add a higher percentage of green ingredients when you start the hot process. If you leave it and don't treat it, it will take longer to break down. Now you have your first layer ready for depositing food scraps 1/week for up to 14 weeks!

Your bottom layer will have 2 rows of 4 deposits each. The next layer will have 3 deposits in a row, then 2 deposits on the next layer, then 1 deposit on the top. This will make a pyramid shape. When you make your first deposit, bury it about a foot in from the sides. (This gives a buffer for anaerobic gas fixation by the aerobic layer and helps to prevent animals from getting into it, however you may still need to have fencing). Mark this deposit. The following deposit is buried next to the first, place your marker in the new spot. When you have 8 deposits for the base, layer on a mixture of green and woody material a bit higher than your deposit height would be. Make 3 deposits in this layer, then layer more green/woody and start another layer with 2 deposits, then the final one in the top center. Cover that and wait 2 weeks. Water layers throughout the process so that the layered materials remain moist.

If after 2 weeks when you open up the pile it has really strong smells then wait another week and open it up again to test. When done, if you plan to hot process it then mix up the whole pile. You can chop up the bottom layer to add extra fungal growth or just leave the base in place. Then measure out how much material you have and start to formulate your hot compost pile recipe. To begin a thermophilic compost pile for your static material, add roughly 10% of very high nitrogen ingredients to the total gallons of the static pile. So if that pile is 200 gallons, add 20 gallons of a high nitrogen ingredient. If you didn't add much, if any, greens consider adding up to 30% greens to the total gallons since the pile may have lost nitrogen over time. This way heat will be maintained throughout when the hot process begins.

Natural Static Pile

Another variation of a static pile is to collect natural ingredients from your surroundings as the woody/green mix. First, soak the woody material in water overnight. Then pile it all together and add to it over time. When you want to stop, give it a good covering with hay or wood chips. Aim to keep it around 50% moisture.

After the static pile breaks down, the material can be used for adding local, biodiverse organisms as part of your woody ingredient for a hot pile process. Alternatively, it can be left for nature to enjoy which will improve your microclimate, especially by planting in it someday.

Static Pile/Hot Pile/Vermicompost vs. Burn Piles/BioChar/BioDigesters

Many people burn throughout the winter season to get rid of backyard 'waste' and to reduce the fuel hazard here in the California foothills. A better option would be for some of that material to be used to make natural static piles, which can hold water from the rains and create more moisture in the air. With enough people, farms, and organizations (i.e. the Forest Service and Cal Fire) making strategic water-retaining static piles, rather than burning, we could have cleaner air. We get enough smoke from the wildfires in the summer! We could also have a restored small-water cycle which is supposed to be the majority of a region's rains.

Additionally, burning wood chips and tree trimmings for biochar has gained a lot of popularity, but it completely kills everything in the material. You still have to add all the biology back into it for it to become biochar. *By burning, you lose out on maintaining organism populations which hydrate the landscape. Decaying local wood builds fungal matter which will ultimately help water to last throughout the dry season.*

On a commercial scale, a bio-digester is a management option that is better than the outdated landfill methods. In some circumstances it may be a better choice for processing large amounts of food waste at once, as well as the added fuel byproduct benefit. However, they are costly and due to extreme heat the organisms die and diverse aerobes have to be added to the end compost product. Current commercial composting facilities that use large compost turners don't account for the required temperature/time threshold which makes high enough populations of diverse, beneficial microbes to treat the soil. Our regulations haven't caught up with this need. Adding biology to commercial compost can be done with applications of thermophilic compost as an amendment to build back the soil biome over time. Another option would be treating it through a large-scale vermicompost process.

Vermicomposting

There are many ways to keep worms for vermicomposting. You can discover and research the types online to find a method that suits your needs. With any variation of worm container you use, they need to be kept out of direct sunlight.

The most popular variety of worm is the red wigglers. The red wiggler worms' optimum temperature range is 70° to 80°. Above 95° or below 39°, you risk killing your worms. By keeping a 70/30 bedding to food waste ratio, the worms' habitat will be aerated and pH neutral. Another method for feeding worms is to solely add composted material instead of food scraps to your worm habitat. This works because the worms are actually eating the organisms of a material, not the food scraps themselves. The worms will increase the beneficial fungal matter and ensure the end product is pathogen-free.

You have good moisture when you squeeze a clump of soil and a drop of liquid comes out. You could use a moisture meter, but are known to be inaccurate. Use a spray bottle for adding water.

Fruit flies and ants are good decomposers but if they get out of hand then it needs some troubleshooting. If there are too many fruit flies, then the worm habitat is too moist and/or there is too much food for worms to process. Stop feeding them for a few days and add more bedding. If there are too many ants, then the worm habitat could be too dry.

There are strict guidelines for what to feed and what not to feed your worms. The Urban Worm Company has a manual containing a quick reference chart and can be downloaded at **urbanwormcompany.com.** They are the makers of the Urban Worm Bag. There are many other worm food lists you can find online to print, or posters or magnets to buy, and put up for handy reminders.

An alternative worm species to have are the nightcrawlers, who have different requirements, but are also very effective for cool processing. The common earthworm is not typically used in a container as they don't process as fast as the red wigglers or nightcrawlers. They live deeper in the soil but will be attracted to your aging compost to provide their benefits.

More On Composting

Dealing with compost materials can be a kind of chicken and egg issue. It is best to have ample materials and clean water that contain the beneficial organisms first. Populate them in order to deal with a material and/or water that is saturated with toxins. So while you can treat toxic environments with biological compost, you have to have enough of the good biology first in order to make it.

You may be wondering what you can and can't put in your compost pile and this depends on what kind of pile you are building. For instance, to compost a dead animal, a 30:1 C:N ratio, it is a single deposit in a static pile. Wait up to 9 months or more for it to break down, then treat for pathogens by using the material in a hot pile process. Or, analogous to the story of 'Where the Red Fern Grows', leave it in the static pile for nature to continue processing it.

Generally speaking, by following the hot pile process (consisting of the 10% high nitrogen, 30% greens, and 60% browns, as well as the appropriate temperature/time threshold turns) you are feeding and populating beneficial microbes. By using a few handfuls of your final compost as starter material you can ensure good biology for the new pile. Such a diverse product makes it possible for these organisms to decompose toxins. Heat does take care of pathogens but a material with too many pathogens won't have the nutrients to feed the aerobes and likely won't get hot enough long enough through the hot process. If ingredients aren't fresh, many of the nutrients are lost so what was a green or high nitrogen material may have to be used as a woody material.

Animal manure should be free of antibiotics. Either obtain manure from a care-taker who doesn't administer antibiotics, or make sure that at least a week has gone by since the animal had any. If you are going to try to use anything with toxic materials, you must increase the size of your pile by adding green and woody material.

If you use an ingredient that has antibiotics in it, find materials for feeding microbes that specifically contain pseudomonas bacteria. Most woody fungal foods take care of the antifungal barriers. Hormones are mostly a nitrogen source so they will get taken up rapidly by microbes. Materials containing synthetic herbicides, let alone pesticides, will kill your organisms. Charcoal (including bio-char) has no nutrition in it to feed microbes in the pile. The coals are 'empty houses' for the microbes and are not used for a hot pile. Ash and other fine ingredients can cause too much of an anaerobic environment because it will compact easily and smother out the aerobes.

Tea or extract that goes bad can be poured over an aging wood chip pile where the humic acids they contain can fix, or neutralize, it. Also, another option for treating chlorine and chloramine in your water is to add a drop of humic acid/per gallon of water and this will neutralize those chemicals.

There are endless scientific nuances in dealing with different types of materials as well as tools being sold for all the steps throughout the process. Visit our website to get on our email list or visit us on Instagram to stay in touch. Other compost and soil consultants in your area can also be found on the Soil Food Web School's website.

This **'Make Soil Not War'** Compost Record book may be the start of your microbe-rich journey and possibly inspire you to scale up your operation. You could get a compost turner (check out **ecoverse.net**) and begin to provide a high quality product for your region!

Compost vs. Tea vs. Extract

Finished compost can be used as straight compost, compost tea, and compost extract.

Using compost straight up: Put your compost around your plants to amend the soil. This behaves similar to the broken down organic matter, or humus, that nature makes. Cover with mulch to protect the moisture of the compost.

Using as compost tea: A tea is meant to be applied as a foliar spray to cover all leaf and stem surfaces of the plant. The brewing process gets all the microbes off the substrate and into the water. Then when it is applied as a spray the organisms stick to the surfaces of the plant. Do this by placing a specified amount of compost in a 400 micro-mesh filter bag and suspend it in water in a brewer tank. There, a tea is brewed for 24 hours with a source of oxygen via an oxygen pump. This is also a chance to add foods that can increase populations of the microbes, such as fungal foods for fungi. This tea lasts only 24 hours before it needs to be used up. Recipes and instructions can be widely found online.

Using as compost extract: Extracts are also a liquid application but are made simply by placing compost in water and stirring for at least 10 minutes. A technical use of an extract is to inject it down into the root zone of the plant with an injection tool. This will break up compaction and get the microbes down deeper so that roots will grow downward instead of competing sideways with other plants and the plant not growing to its potential. On a big and technical scale you can make an extract in a large vessel and connect it to a hose which connects to an injector tool. This is applied halfway between the stem and drip line of the plant. On a smaller scale place a handful of compost in a 5-gallon bucket of water and apply it using a watering can. Extracts also only last 24 hours.

This compost extract will also increase your fungi to bacteria ratio for your next compost pile. Soak your woody material in extract for 2 days.

HOW TO USE

How To Use Record Book Pages

Give yourself time to read through all the sections of this book first before getting started. It is a lot of information to take in and will take time to be fully prepared for your composting voyage.

Scratch Recipe Page

The Record Book section begins with the scratch recipe page for figuring out your compost ingredient quantities. Figuring out your ratios of ingredients is a little bit like algebra. You may have to work with what you get, like coffee grounds from a café, or whatever spent grain someone is willing to share etc., so you could start by measuring what you have collected into gallons.

For example, if you pick up 5 gallons of coffee grounds, how much more greens do you need for a 30% greens ratio in a 150 gallon pile?

150 gallons times .3 = 45 gallons green. Minus the 5 gallons you already have of coffee grounds means you will need 40 gallons more green material.

What if you get 20 gallons of a high nitrogen material and you want to use it all, but you also need diversity so you want to add a few gallons of some other high nitrogen material. This would then make an over 200 gallon pile, and you would need to make sure you have enough of the other ingredients to successfully achieve the 10%, 30%, 60% ratio. Remember something like spent grains from a brewery are already soaking, they will only last a day before they start going anaerobic and begin off-gassing and losing nutrients.

Compost Pile # / Final Compost Recipe Page

Next, transfer what you came up with from the Scratch Recipe page into an organized, easy to read format in the Final Compost Recipe table.

Calendar Pages

Start using the calendar pages when you are ready to start tracking your pile. It can be started on any month on any day. Your pile could take up to 15 days, and another 10 days to cool, sometimes more or less depending on how well you can predict how a particular pile will behave. If you plan to be out of town, use the calendar pages to help you plan your schedule to avoid conflict with the pile process.

Examples for Notes: what is your target successional stage; when and where you got your ingredients; details of starting material, like fresh or dried and rehydrated.

HOW TO USE

Here is a checklist of notes to make on your calendar:
- The 10 day Forecast (to prepare for rain, wind)
- When you started your pile and when it got to 131°
- When your pile ended the thermophilic phase
- When your pile got to ambient temperatures
- When you took soil samples
- When you received your soil/compost report data

Daily Observation Log

Next are the pages titled Daily Observation Log where you will start to log data. When the temperature of your hot compost pile hits 131°, this will be day 0. Make a note for how many days it took to get up to that temperature in the notes section.

Use the Time Interval Box for writing in the time period as you go. After day 1 you would write 24 for 24 hours, after day 2 write 48 hours and so on. This will help you keep track of the hours for the heat cycle you are in and anticipate which day and time you need to turn your pile. Follow across and fill out the ambient temperature, and the average temperature and moisture level. For the Threshold Reached Column, write what threshold you are working in, either 131°, 150°, or 160°.

On turning day the boxes labeled Day, Time, and Threshold Reached are shared between the last day of a cycle and the first day of a cycle, so make split(s) in the boxes. The average of the 3 temperatures and moistures will also need to be shared.

When you turn your pile mark a T in the Day box as well as the day. In the Time box make an additional note of the time when the pile hits 131°+ again. This is the new start time for the pile. In the threshold reached box note the new threshold temperature, as well as average temp. and moisture.

You will also need to share boxes if at any time during the cycle, you observe that the time/temp. threshold has changed. Update the time you observed the temperature and make sure the hours correspond with this in the time interval box. The next day at this new time will be 24 hours for the new temp. threshold reached.

After the thermophilic process has completed, you will still have to monitor moisture and measure temperature to know when the pile has cooled to outside ambient temperatures. You can track these details in the log and transfer to the summary page later.

*Separate additional Observation Log pages can be made available on request, email **contact@soilterroir.com**.

HOW TO USE

How To Measure Temperature

After creating or turning your pile, wait at least an hour before taking the temperature. The goal here is to get the temperature of the center of the pile from three different entry points, and record the average in the daily observation log pages. To get the average of three readings you will take one reading from the top of the pile and two from the sides (see steps and illustrations). Use the scratch part of your book to note each reading and record the average of the three in the daily observation log table.

Obtaining a reading from the top of a pile:

Step 1: Place the thermometer next to the pile, lifting or lowering if you need to so that the head is level with the top of the pile contents. Grab and hold the point of the thermometer's stem that is at the vertical halfway point of the pile, this is marking how far down you will be going in from the top.

Step 2: Then insert the thermometer in at the top center of the pile and go down up to the point where your fingers are holding it from step one. Remember this temperature reading, or jot it down.

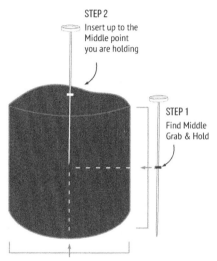

Obtaining a reading from the sides of the pile:

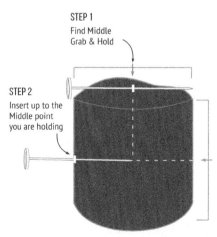

Step 1: Lay your thermometer across the top, directly over the center, with the head of it just outside the vessel. Grab and hold the point of the thermometer where it looks most lined up in the middle, this is marking how far the thermometer is going in from the side.

Step 2: Eyeing a point on the side of the pile that is halfway between the top of the compost material and the bottom of the pile, slide the thermometer in until where your fingers are holding the spot from step one. Repeat these steps from another point of entry for your 3rd reading. Now take the average of the 2 sides and the 1 top reading to record temperature.

How to Calibrate Your Thermometer

Every few months you need to calibrate your thermometer for accuracy.

Step 1: Check to see if your thermometer reads 32° F, or 0° C, by inserting the tip into ice water. If after a minute or so it does not then read the thermometer's manual on how to take apart the dial. Once the dial is apart, follow the manual instructions to calibrate it for freezing.

Step 2: Upon getting a 32° F reading, you then check for the boiling point at 212° F or 100° C. Insert the tip of the thermometer while the water is actively boiling. If your thermometer doesn't have a 212° F (100° C) reading, it will go to the end to signify boiling. If it doesn't get to 212° F (100°C), or doesn't go to the end then you will want to repeat step 1 following the instructions in the thermometer's manual, and then step 2 again. Notes on thermometer: A 3' thermometer is ideal for a pile whose vessel is 3' high by 3' across the top. The pile can be bigger but the thermometer just has to be big enough that its endpoint reaches the center of the hot middle part of the pile. The Reo thermometer is a trusted brand.

How To Measure Moisture

Here is where the rubber gloves come in, but you may or may not choose to wear them. Reach into the pile up to your elbows, or as far as the rubber gloves go, pull out a little handful of compost and squeeze it. If there are several drops between your fingers, you have over 50% moisture. If there is basically 1 drop between your fingers that is about 50%, and no moisture coming out means it is below 50%. Do this in 3 places and record the average. Remember too much moisture can cause compaction, both on compost and on bare soil. Too little will inhibit organism growth and the pile won't heat up. (This moisture test does not apply to the static pile or the vermicompost process.) Once cooled, the moisture needs to be maintained at 35%. This is observed when a handful holds together but doesn't drip.

Biological Assessments

After your pile has cooled, you can assess the pile's biology by getting a soil test. The Biological Assessment Report Results page is a space to record your soil biology summary upon getting soil test results. Fungi increase over time so you may want to wait a month or two to get your pile tested. You could also record report data of your garden soil for each season on the Season Summary Notes pages.

How To Order A Biological Assessment Report

A biological soil test will tell you the quantifications of organisms in your compost and soil. It will help you to manage the number and types of applications throughout the year. A test also demonstrates when your compost should be applied so that it more closely matches the successional stage of the plants you are growing. Make sure the biological test that you order performs the shadow microscopy technique to ensure you meet the minimum requirements for how, when, and where you want to apply the compost.

Soil tests are best as fresh as possible so locate a soil microscopist in your area. Visit the Soil Food Web website's consultants page for one closest to you. Soil Terroir is based in Grass Valley, California and offers Biological Microscopy Assessments of Soil and Compost. They do assessments of compost tea and extract by hand delivery method only. Sample collection protocol is described in detail at **SoilTerroir.com/faq**

Once your assessment has been purchased via the Soil Terroir website, arrangements by hand delivery or shipping can be done. If you are within 20 miles of Grass Valley, California, free drop off and pick up is available. To send in your sample(s), package them in a small box or padded envelope to avoid breaking the seal of the bag and compaction of the sample. For accurate assessment results, samples must arrive as quickly as possible. For example, during cooler months, a 2-3 day arrival window is sufficient for mailing solid samples. For warmer months, next day delivery is recommended. Also make sure to ship your sample so that it arrives during the weekdays of Monday – Thursday. For more information, visit **SoilTerroir.com** or email Hilary at **contact@soilterroir.com.**

Understanding a Biological Soil Report

Since optimum growth of a plant is shown when the ratio of beneficial bacteria to fungi is at the right balance for a plant's successional stage, then it makes sense to quantify the bacteria and fungi as a ratio in a soil report. The fungi and bacteria are comparable in mass and are measured as such because what bacteria lack in size, they make up for in number. Conversely, fungi are larger but there are not as many. For example, the largest organism on the planet is a single fungi, which is 1,665 football fields and resides in Oregon, USA. To quantify them and get an accurate comparison we convert to a measurement that works for both in micrograms/gram (μg/g). This is considered their biomass.

Here is a quick rundown of an example soil report interpretation:

In a report we look at the amount of fungi first. If fungi is low, the standard deviation will also be low, and this just means that you need to increase the amount of fungi, which it states on the report.

Then we look at the Fungi to Bacteria (F:B) ratio. If there is too little fungi compared to bacteria, it will read 0. Next we look at bacteria. If it is high, increase the fungi and make sure you have ample predators.

Next we look at protozoa. Make sure the numbers are sufficient for all the bacteria. For standard deviation, you want it to be less than a third of the total population.

For nematodes, you won't see any fungal feeders if there are no fungi. Later in succession, you will not see as many bacterial feeders as they are replaced with an increase in micro-arthropods. It is ok to have ciliates as long as amoebae and flagellates are adequate in number.

Garden Map & Reflection Pages

This space is for drawing your garden to visualize and record which plants will get amendments and when. Once you draw your map, mark where you added compost, tea, or extract. Use accompanying garden map notes for more details and plans.

Appendix Pages

Static Pile Log
Included in this static pile record is a column for the deposit material and size of it. However, if you don't deposit anything then you may just leave that blank. You can creatively use this log to record whatever static pile you choose to make, it would be curious to see what it looks like under a microscope when it breaks down and compare between different piles. You will also find Daily Observation Log pages after each static pile if you decide to process the materials thermophilically.

Worm Bin Log
The worm bin log pages are for documenting its details from the start, suggested on a weekly basis, for about a year.

Season Summary Pages
As you apply your compost to your garden, you can use these pages to make note of how your garden is growing each season. Should you get a soil test, you can track improvement of your soil over time and ensure it is complete with all the functioning organism groups for your plants. Your plants will thank you!

Q & A Notes
This section is for any questions that come up during this process that you plan to research, consult with someone, or ask in a workshop/webinar etc.

Lists Of References & Educational Notes
These pages are for lists of references, or things you want to do; such as a book list, webinars or workshops to attend. Additional pages can be used for note-taking.

Resources & Sourcing Contacts
Collecting your compost making ingredients takes time. Along the way you may secure sources of your materials. Here you can keep your contacts organized for the next time you need more ingredients. For woodchips contacts, write the date you obtain woodchips in the notes so you know they are at least six months old.

THE COMPOST RECORD BOOK

Configure Your Compost Recipe:

Compost Pile #1

Start Date _____

End Date _____

Final Compost Recipe	
STARTING MATERIALS	BUCKETS
HIGH N	10%
GREEN	30%
WOODY	60%
TOTAL BUCKETS	

Calendar

Notes

Daily Observation Log

TIME INTERVAL	DAY / TURN	DATE	TIME	THRESHOLD REACHED	AMBIENT TEMP
☐					
☐					
☐					
☐					
☐					
☐					
☐					
☐					
☐					

Daily Observation Log

AVERAGE OF 3 TEMPS	AVERAGE OF 3 MOISTURES	NOTES

Daily Observation Log

TIME INTERVAL	DAY / TURN	DATE	TIME	THRESHOLD REACHED	AMBIENT TEMP

Daily Observation Log

AVERAGE OF 3 TEMPS	AVERAGE OF 3 MOISTURES	NOTES

Compost Pile Summary

BIOLOGICAL ASSESSMENT REPORT RESULTS

BIOLOGY		
Bacterial Biomass		
Actinobacterial Biomass		
Fungal Biomass		
F:B Ratio		
Flagellates		
Amoebae		
Nematodes		
Bacterial-Feeding Nematodes		
Fungal-feeding Nematodes		
Predatory Nematodes		
Root-Feeding Nematodes		
Oomycetes Biomass		
Ciliates		
Other Notes:		

Compost Pile Summary

HEAT CYCLE SUMMARY LOG

COMPOST	DATE THRESHOLD REACHED/PILE TURNED	TEMP THRESHOLD	# OF HOURS	MAX TEMP
HEAT CYCLE 1				
HEAT CYCLE 2				
HEAT CYCLE 3				
SUMMARY	NUMBER OF DAYS	AMBIENT TEMPERATURES	AVERAGE MOISTURE	COVER (TARP, SHEET, ETC?)
THERMOPHILIC PHASE:				
MATURATION PHASE:				

Draw Your Garden Map

Lessons Learned From Pile #1

Where, When & How To Use Pile #1

Configure Your Compost Recipe:

Compost Pile #2

Start Date _____

End Date _____

Final Compost Recipe	
STARTING MATERIALS	BUCKETS
HIGH N	10%
GREEN	30%
WOODY	60%
TOTAL BUCKETS	

Calendar

Notes

Daily Observation Log

TIME INTERVAL	DAY / TURN	DATE	TIME	THRESHOLD REACHED	AMBIENT TEMP

Daily Observation Log

AVERAGE OF 3 TEMPS	AVERAGE OF 3 MOISTURES	NOTES

Daily Observation Log

TIME INTERVAL	DAY / TURN	DATE	TIME	THRESHOLD REACHED	AMBIENT TEMP

Daily Observation Log

AVERAGE OF 3 TEMPS	AVERAGE OF 3 MOISTURES	NOTES

Compost Pile Summary

BIOLOGICAL ASSESSMENT REPORT RESULTS

BIOLOGY		
Bacterial Biomass		
Actinobacterial Biomass		
Fungal Biomass		
F:B Ratio		
Flagellates		
Amoebae		
Nematodes		
Bacterial-Feeding Nematodes		
Fungal-feeding Nematodes		
Predatory Nematodes		
Root-Feeding Nematodes		
Oomycetes Biomass		
Ciliates		
Other Notes:		

Compost Pile Summary

HEAT CYCLE SUMMARY LOG

COMPOST	DATE THRESHOLD REACHED/PILE TURNED	TEMP THRESHOLD	# OF HOURS	MAX TEMP
HEAT CYCLE 1				
HEAT CYCLE 2				
HEAT CYCLE 3				
SUMMARY	NUMBER OF DAYS	AMBIENT TEMPERATURES	AVERAGE MOISTURE	COVER (TARP, SHEET, ETC?)
THERMOPHILIC PHASE:				
MATURATION PHASE:				

Draw Your Garden Map

Lessons Learned From Pile #2

Where, When & How To Use Pile #2

Configure Your Compost Recipe:

Compost Pile #3

Start Date _____

End Date _____

Final Compost Recipe	
STARTING MATERIALS	BUCKETS
HIGH N	10%
GREEN	30%
WOODY	60%
TOTAL BUCKETS	

Calendar

Notes

Daily Observation Log

TIME INTERVAL	DAY / TURN	DATE	TIME	THRESHOLD REACHED	AMBIENT TEMP
☐					
☐					
☐					
☐					
☐					
☐					
☐					
☐					
☐					

Daily Observation Log

AVERAGE OF 3 TEMPS	AVERAGE OF 3 MOISTURES	NOTES

Daily Observation Log

TIME INTERVAL	DAY / TURN	DATE	TIME	THRESHOLD REACHED	AMBIENT TEMP

Daily Observation Log

AVERAGE OF 3 TEMPS	AVERAGE OF 3 MOISTURES	NOTES

Compost Pile Summary

BIOLOGICAL ASSESSMENT REPORT RESULTS

BIOLOGY		
Bacterial Biomass		
Actinobacterial Biomass		
Fungal Biomass		
F:B Ratio		
Flagellates		
Amoebae		
Nematodes		
Bacterial-Feeding Nematodes		
Fungal-feeding Nematodes		
Predatory Nematodes		
Root-Feeding Nematodes		
Oomycetes Biomass		
Ciliates		
Other Notes:		

Compost Pile Summary

HEAT CYCLE SUMMARY LOG

COMPOST	DATE THRESHOLD REACHED/PILE TURNED	TEMP THRESHOLD	# OF HOURS	MAX TEMP
HEAT CYCLE 1				
HEAT CYCLE 2				
HEAT CYCLE 3				
SUMMARY	NUMBER OF DAYS	AMBIENT TEMPERATURES	AVERAGE MOISTURE	COVER (TARP, SHEET, ETC?)
THERMOPHILIC PHASE:				
MATURATION PHASE:				

Draw Your Garden Map

Lessons Learned From Pile #3

Where, When & How To Use Pile #3

Configure Your Compost Recipe:

Compost Pile #4

Start Date _____

End Date _____

Final Compost Recipe	
STARTING MATERIALS	BUCKETS
HIGH N	10%
GREEN	30%
WOODY	60%
TOTAL BUCKETS	

Calendar

Notes

Daily Observation Log

TIME INTERVAL	DAY / TURN	DATE	TIME	THRESHOLD REACHED	AMBIENT TEMP

Daily Observation Log

AVERAGE OF 3 TEMPS	AVERAGE OF 3 MOISTURES	NOTES

Daily Observation Log

TIME INTERVAL	DAY / TURN	DATE	TIME	THRESHOLD REACHED	AMBIENT TEMP
☐					
☐					
☐					
☐					
☐					
☐					
☐					
☐					
☐					

Daily Observation Log

AVERAGE OF 3 TEMPS	AVERAGE OF 3 MOISTURES	NOTES

Compost Pile Summary

BIOLOGICAL ASSESSMENT REPORT RESULTS

BIOLOGY		
Bacterial Biomass		
Actinobacterial Biomass		
Fungal Biomass		
F:B Ratio		
Flagellates		
Amoebae		
Nematodes		
Bacterial-Feeding Nematodes		
Fungal-feeding Nematodes		
Predatory Nematodes		
Root-Feeding Nematodes		
Oomycetes Biomass		
Ciliates		
Other Notes:		

Compost Pile Summary

HEAT CYCLE SUMMARY LOG

COMPOST	DATE THRESHOLD REACHED/PILE TURNED	TEMP THRESHOLD	# OF HOURS	MAX TEMP
HEAT CYCLE 1				
HEAT CYCLE 2				
HEAT CYCLE 3				
SUMMARY	NUMBER OF DAYS	AMBIENT TEMPERATURES	AVERAGE MOISTURE	COVER (TARP, SHEET, ETC?)
THERMOPHILIC PHASE:				
MATURATION PHASE:				

Draw Your Garden Map

Lessons Learned From Pile #4

Where, When & How To Use Pile #4

Configure Your Compost Recipe:

Compost Pile #5

Start Date _____

End Date _____

Final Compost Recipe	
STARTING MATERIALS	BUCKETS
HIGH N	10%
GREEN	30%
WOODY	60%
TOTAL BUCKETS	

Calendar

Notes

Daily Observation Log

TIME INTERVAL	DAY / TURN	DATE	TIME	THRESHOLD REACHED	AMBIENT TEMP
☐					
☐					
☐					
☐					
☐					
☐					
☐					
☐					
☐					
☐					

Daily Observation Log

AVERAGE OF 3 TEMPS	AVERAGE OF 3 MOISTURES	NOTES

Daily Observation Log

TIME INTERVAL	DAY / TURN	DATE	TIME	THRESHOLD REACHED	AMBIENT TEMP

Daily Observation Log

AVERAGE OF 3 TEMPS	AVERAGE OF 3 MOISTURES	NOTES

Compost Pile Summary

BIOLOGICAL ASSESSMENT REPORT RESULTS

BIOLOGY		
Bacterial Biomass		
Actinobacterial Biomass		
Fungal Biomass		
F:B Ratio		
Flagellates		
Amoebae		
Nematodes		
Bacterial-Feeding Nematodes		
Fungal-feeding Nematodes		
Predatory Nematodes		
Root-Feeding Nematodes		
Oomycetes Biomass		
Ciliates		
Other Notes:		

Compost Pile Summary

HEAT CYCLE SUMMARY LOG

COMPOST	DATE THRESHOLD REACHED/PILE TURNED	TEMP THRESHOLD	# OF HOURS	MAX TEMP
HEAT CYCLE 1				
HEAT CYCLE 2				
HEAT CYCLE 3				
SUMMARY	NUMBER OF DAYS	AMBIENT TEMPERATURES	AVERAGE MOISTURE	COVER (TARP, SHEET, ETC?)
THERMOPHILIC PHASE:				
MATURATION PHASE:				

Draw Your Garden Map

Lessons Learned From Pile #5

Where, When & How To Use Pile #5

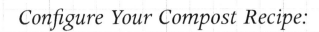

Compost Pile #6

Start Date _____

End Date _____

Final Compost Recipe	
STARTING MATERIALS	BUCKETS
HIGH N	10%
GREEN	30%
WOODY	60%
TOTAL BUCKETS	

Calendar

Notes

Daily Observation Log

TIME INTERVAL	DAY / TURN	DATE	TIME	THRESHOLD REACHED	AMBIENT TEMP

Daily Observation Log

AVERAGE OF 3 TEMPS	AVERAGE OF 3 MOISTURES	NOTES

Daily Observation Log

TIME INTERVAL	DAY / TURN	DATE	TIME	THRESHOLD REACHED	AMBIENT TEMP
☐					
☐					
☐					
☐					
☐					
☐					
☐					
☐					
☐					

Daily Observation Log

AVERAGE OF 3 TEMPS	AVERAGE OF 3 MOISTURES	NOTES

Compost Pile Summary

BIOLOGICAL ASSESSMENT REPORT RESULTS

BIOLOGY		
Bacterial Biomass		
Actinobacterial Biomass		
Fungal Biomass		
F:B Ratio		
Flagellates		
Amoebae		
Nematodes		
Bacterial-Feeding Nematodes		
Fungal-feeding Nematodes		
Predatory Nematodes		
Root-Feeding Nematodes		
Oomycetes Biomass		
Ciliates		
Other Notes:		

Compost Pile Summary

HEAT CYCLE SUMMARY LOG

COMPOST	DATE THRESHOLD REACHED/PILE TURNED	TEMP THRESHOLD	# OF HOURS	MAX TEMP
HEAT CYCLE 1				
HEAT CYCLE 2				
HEAT CYCLE 3				
SUMMARY	NUMBER OF DAYS	AMBIENT TEMPERATURES	AVERAGE MOISTURE	COVER (TARP, SHEET, ETC?)
THERMOPHILIC PHASE:				
MATURATION PHASE:				

Draw Your Garden Map

Lessons Learned From Pile #6

Where, When & How To Use Pile #6

APPENDIX

Static Pile Log #1

Season: _____ Size of Base: _____

Water Source: _____   _____

DATE	DEPOSIT MATERIAL	SIZE OF DEPOSIT	TYPE OF MATERIAL ADDED TO COVERGREEN AND/OR WOODY ?	AMOUNT OF MATERIAL

Static Pile Log

Weekly Deposit Estimate: _____

Weekly Material Estimate: _____

WATERED Y/N	NOTES

Daily Observation Log #1

TIME INTERVAL	DAY / TURN	DATE	TIME	THRESHOLD REACHED	AMBIENT TEMP
☐					
☐					
☐					
☐					
☐					
☐					
☐					
☐					
☐					
☐					

Daily Observation Log

AVERAGE OF 3 TEMPS	AVERAGE OF 3 MOISTURES	NOTES

Daily Observation Log #1

TIME INTERVAL	DAY / TURN	DATE	TIME	THRESHOLD REACHED	AMBIENT TEMP

Daily Observation Log

AVERAGE OF 3 TEMPS	AVERAGE OF 3 MOISTURES	NOTES

Static Pile Log #2

Season: _____   Size of Base: _____

Water Source: _____   _____

DATE	DEPOSIT MATERIAL	SIZE OF DEPOSIT	TYPE OF MATERIAL ADDED TO COVER GREEN AND/OR WOODY ?	AMOUNT OF MATERIAL

Static Pile Log

Weekly Deposit Estimate: _____

Weekly Material Estimate: _____

WATERED Y/N	NOTES

Daily Observation Log #1

TIME INTERVAL	DAY / TURN	DATE	TIME	THRESHOLD REACHED	AMBIENT TEMP
☐					
☐					
☐					
☐					
☐					
☐					
☐					
☐					
☐					

Daily Observation Log

AVERAGE OF 3 TEMPS	AVERAGE OF 3 MOISTURES	NOTES

Daily Observation Log #1

TIME INTERVAL	DAY / TURN	DATE	TIME	THRESHOLD REACHED	AMBIENT TEMP

Daily Observation Log

AVERAGE OF 3 TEMPS	AVERAGE OF 3 MOISTURES	NOTES

Worm Bin Log

Worm Species: _____

Worm Source: _____

DATE	GREENS	AMOUNT	BROWNS	AMOUNT

Worm Bin Log

Worm Bin Type:

Bin Location:

MOISTURE %	NOTES

Worm Bin Log

Worm Species: _____

Worm Source: _____

DATE	GREENS	AMOUNT	BROWNS	AMOUNT

Worm Bin Log

Worm Bin Type: _____

Bin Location: _____

MOISTURE %	NOTES

Seasons Summary

Spring

Summer

Seasons Summary

Autumn

Winter

Questions & Answers

Questions & Answers

Lists Of Educational References

Lists Of Educational References

Educational Notes From Workshops/webinars

Educational Notes From Workshops/webinars

Educational Notes

Educational Notes

Resources & Sourcing Contacts

COMPANY
CONTACT NAME
ADDRESS
PHONE
EMAIL
WEBSITE
OTHER NOTES

COMPANY
CONTACT NAME
ADDRESS
PHONE
EMAIL
WEBSITE
OTHER NOTES

COMPANY
CONTACT NAME
ADDRESS
PHONE
EMAIL
WEBSITE
OTHER NOTES

Resources & Sourcing Contacts

COMPANY
CONTACT NAME
ADDRESS
PHONE
EMAIL
WEBSITE
OTHER NOTES

COMPANY
CONTACT NAME
ADDRESS
PHONE
EMAIL
WEBSITE
OTHER NOTES

COMPANY
CONTACT NAME
ADDRESS
PHONE
EMAIL
WEBSITE
OTHER NOTES

Resources & Sourcing Contacts

COMPANY
CONTACT NAME
ADDRESS
PHONE
EMAIL
WEBSITE
OTHER NOTES

COMPANY
CONTACT NAME
ADDRESS
PHONE
EMAIL
WEBSITE
OTHER NOTES

COMPANY
CONTACT NAME
ADDRESS
PHONE
EMAIL
WEBSITE
OTHER NOTES

Resources & Sourcing Contacts

COMPANY
CONTACT NAME
ADDRESS
PHONE
EMAIL
WEBSITE
OTHER NOTES

COMPANY
CONTACT NAME
ADDRESS
PHONE
EMAIL
WEBSITE
OTHER NOTES

COMPANY
CONTACT NAME
ADDRESS
PHONE
EMAIL
WEBSITE
OTHER NOTES

Additional Resources

SOURCES

Soil Food Web School LLC. (2021)
Courses, course manual. Mentor, Adrienne L. Godshalx, PhD
Soilfoodweb.com

Water Stories, Waterstories.com (2021)

Kingdom (biology), Wikipedia, 2022

RECOMMENDED READING

Beyond the War on Invasive Species by Tao Orion
Restoration Agriculture by Mark Shepard
Agriculture by Rudolf Steiner
A Buzz in the Meadow by Dave Goulson
The Living Earth Handbook by Renee Wade

RECOMMENDED WEBSITES

Soil Food Web School soilfoodweb.com
Water Stories waterstories.com
Catalyst BioAmendments catalystbioamendments.com
Urban Worm Company urbanwormcompany.com
Permaculture Women's Guild, Natalie Topa permaculturewomen.com/natalietopa
Didi Pershouse didipershouse.com
Soil Regen Summit soilregensummit.com
Paul Stamets paulstamets.com
Pandora Thomas pandorathomas.com
Practical Permaculture practicalpermaculture.com
Robin Wall Kimmerer robinwallkimmerer.com
Ecoverse ecoverse.net

BIOLOGICAL MICROSCOPY ASSESSMENTS

Soil Terroir (279) 206-3437 | contact@soilterroir.com

Made in the USA
Monee, IL
10 April 2023